**PA**
Somerset County Library System
11767 Beechwood Street
Princess Anne, MD 21853
PRINCESS ANNE BRANCH

## SUPERHERO ANIMALS

# WOLVERINES

**KENNY ABDO**

Fly!
An Imprint of Abdo Zoom
abdobooks.com

**abdobooks.com**

Published by Abdo Zoom, a division of ABDO, P.O. Box 398166, Minneapolis, Minnesota 55439. Copyright © 2020 by Abdo Consulting Group, Inc. International copyrights reserved in all countries. No part of this book may be reproduced in any form without written permission from the publisher. Fly!™ is a trademark and logo of Abdo Zoom.

Printed in the United States of America, North Mankato, Minnesota.
102019
012020

Photo Credits: Alamy, Everett Collection, iStock, Shutterstock, ©Pinguino p8 / CC BY 2.0, ©Lex Larson p8 / CC BY-SA 2.0, ©Gage Skidmore p8 /CC BY-SA 2.0, ©SunGodKizaru Incredible Hulk Vol 1 180 p9 / CC-BY-SA
Production Contributors: Kenny Abdo, Jennie Forsberg, Grace Hansen
Design Contributors: Dorothy Toth, Neil Klinepier

### Library of Congress Control Number: 2019941305

### Publisher's Cataloging-in-Publication Data

Names: Abdo, Kenny, author.
Title: Wolverines / by Kenny Abdo
Description: Minneapolis, Minnesota : Abdo Zoom, 2020 | Series: Superhero animals | Includes online resources and index.
Identifiers: ISBN 9781532129520 (lib. bdg.) | ISBN 9781098220501 (ebook) | ISBN 9781098220990 (Read-to-Me ebook)
Subjects: LCSH: Wolverine--Juvenile literature. | Carcajou--Juvenile literature. | Glutton (Animal)--Juvenile literature. | Mustelids--Juvenile literature. | Scavengers (Zoology)--Juvenile literature. | Zoology--Juvenile literature.
Classification: DDC 599.766--dc23

# TABLE OF CONTENTS

Wolverines . . . . . . . . . . . . . . . . . . . . . . 4

Origin Story . . . . . . . . . . . . . . . . . . . . 8

Powers & Abilities . . . . . . . . . . . . . 12

In Action . . . . . . . . . . . . . . . . . . . . . 18

Glossary . . . . . . . . . . . . . . . . . . . . . 22

Online Resources . . . . . . . . . . . . . 23

Index . . . . . . . . . . . . . . . . . . . . . . . . 24

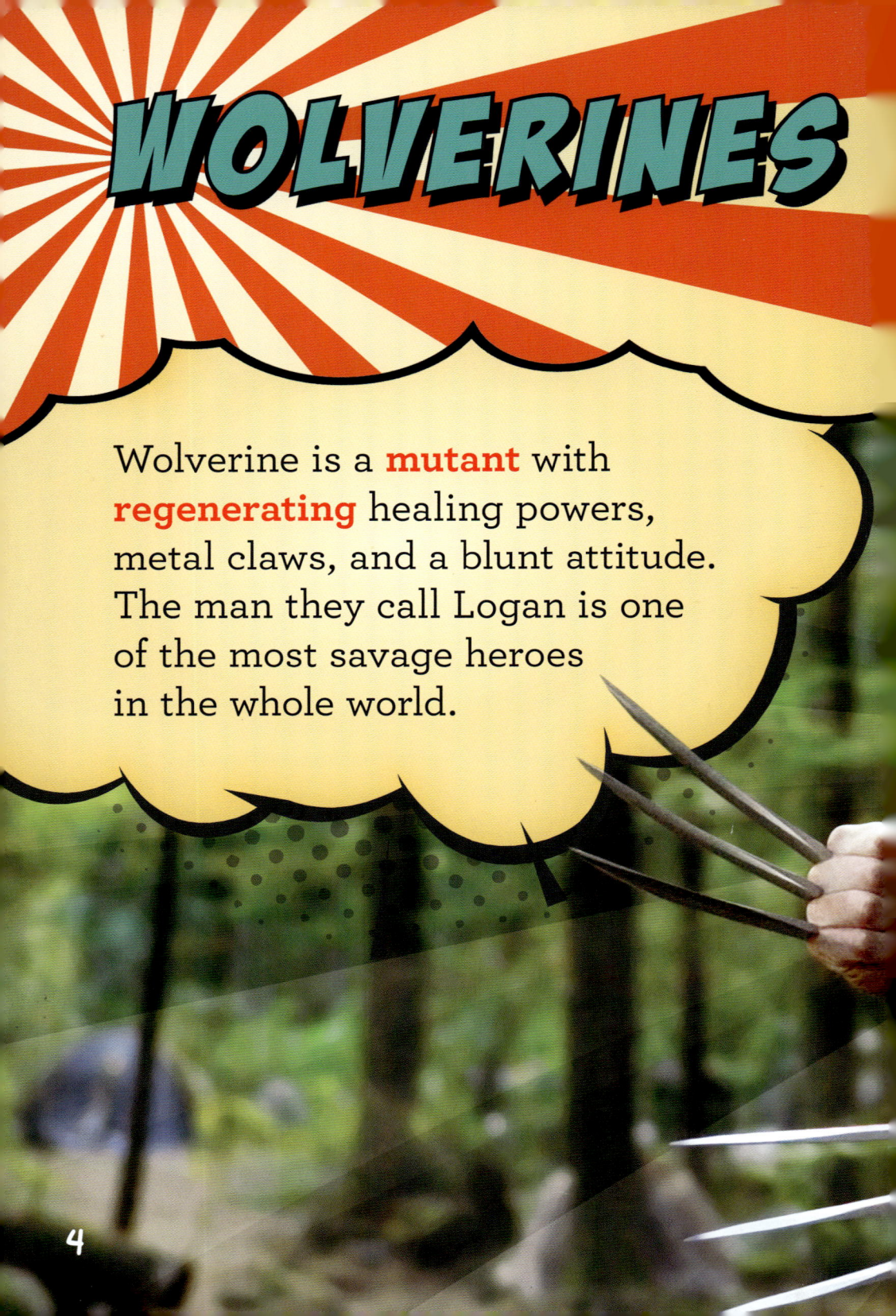

# WOLVERINES

Wolverine is a **mutant** with **regenerating** healing powers, metal claws, and a blunt attitude. The man they call Logan is one of the most savage heroes in the whole world.

The wolverine is a powerful animal with short legs and an even shorter temper.

Wolverine first sliced his way across the page in *The Incredible Hulk* issue 180 in 1974. Roy Thomas and Len Wein came up with the character. Illustrator John Romita Sr. brought Wolverine to life.

The creator's wanted to have a hero who was small in height, but brutal in temper. So, they named him after the animal that best fit that description.

WANNA KNOW WHAT I THINK?

I THINK THE ONLY THING THAT CAN CUT THROUGH ADAMANTIUM...

# POWERS & ABILITIES

Wolverines are **solitary** animals. They prefer colder climates. They build **dens** in the snow.

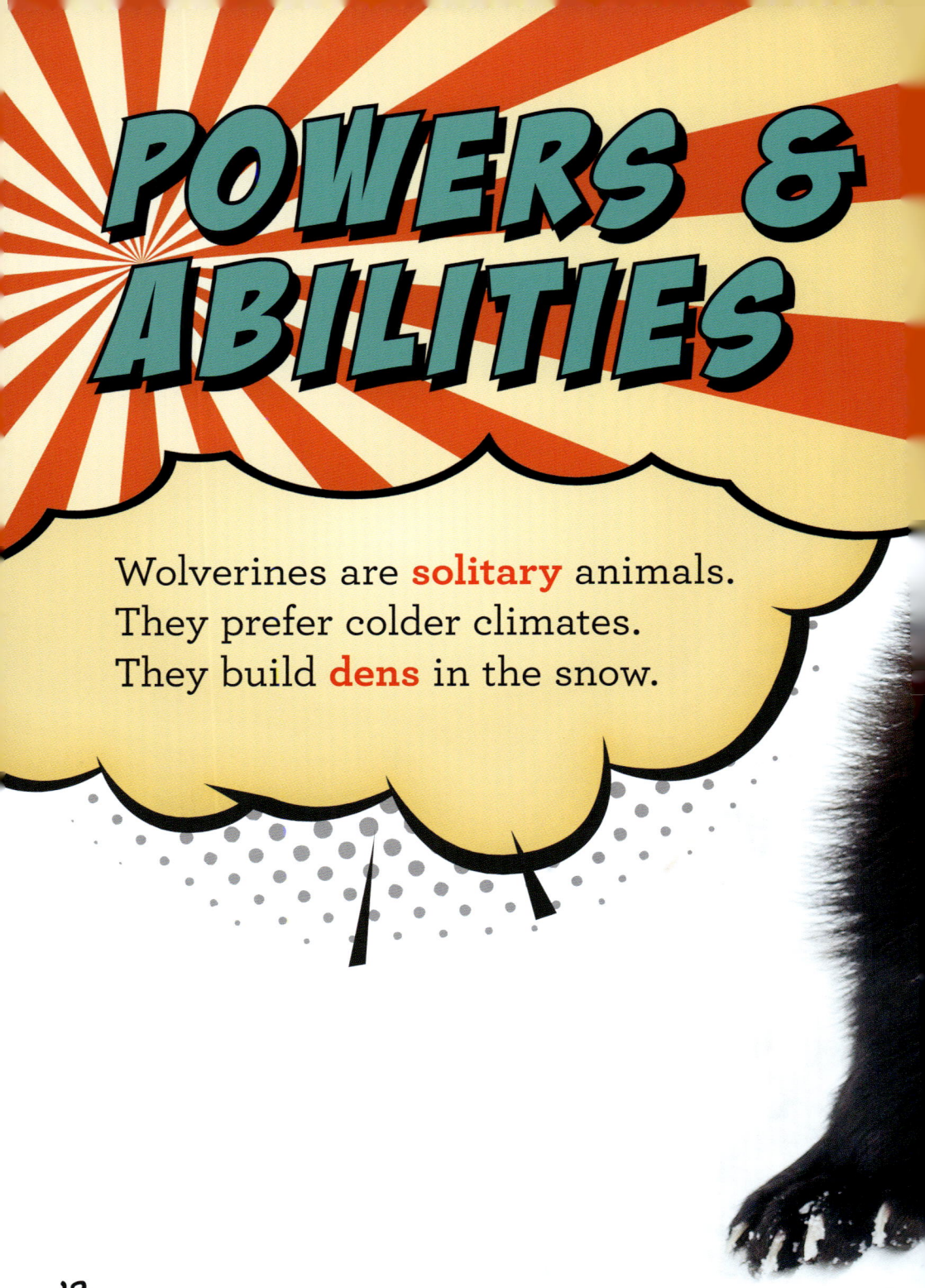

Wolverines are nocturnal. They sleep during the day and hunt at night.

Wolverines have a strong sense of smell. They can smell **prey** buried more than 20 feet (6.1 m) under the snow.

Wolverines have strong jaws and teeth. They are known to eat the bones and teeth of their **prey**.

Wolverines have a thick fur coat that is extremely **hydrophobic**. They are able to rest in the barest of shelters and in the toughest conditions.

# IN ACTION

Wolverine has a heightened sense of smell. It allows him to track people and objects by scent alone.

Aside from **regenerating** healing powers, Wolverine has a **reinforced** skeleton. It is made of **adamantium**. It gives Wolverine superhuman strength.

# INDEX

claws 4, 21

habitat 12

healing 4, 20

*Incredible Hulk, The* (comic) 9

jaws 16

Logan (character) 4

Romita Sr., John 9, 10

smell 15, 18

strength 20

Thomas, Roy 9, 10

Wein, Len 9, 10

Wolverine (character) 4, 9, 10, 18, 20, 21

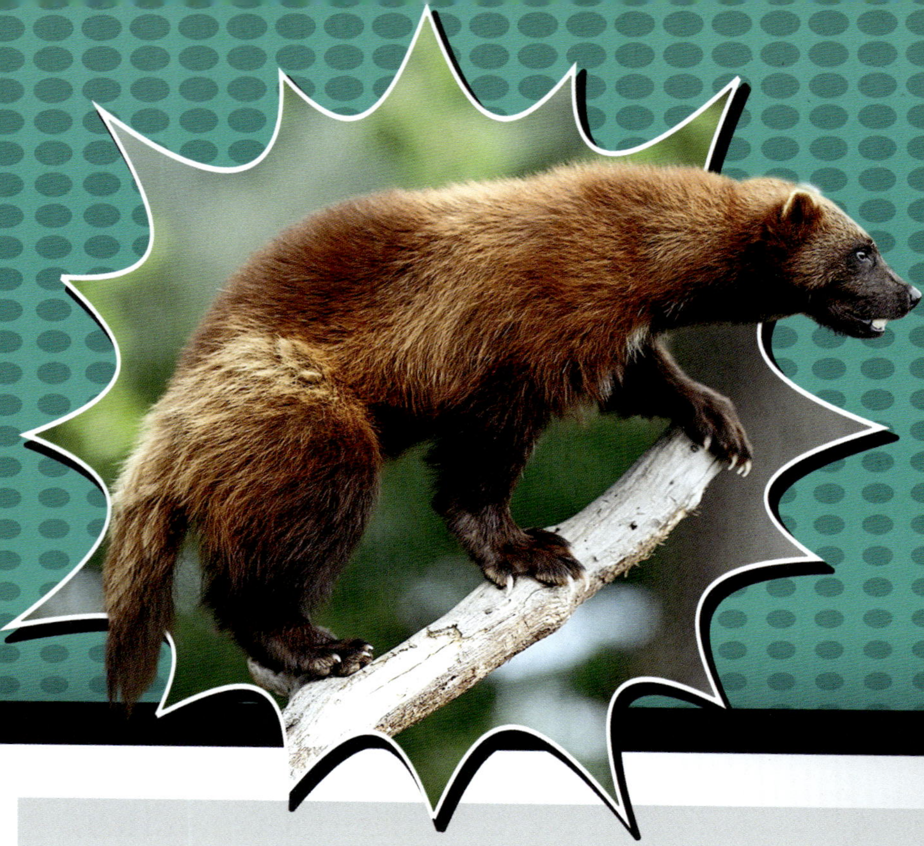

# ONLINE RESOURCES

To learn more about wolverines, please visit **abdobooklinks.com** or scan this QR code. These links are routinely monitored and updated to provide the most current information available.

# GLOSSARY

**adamantium** – a fictional metal that is impossible to destroy.

**den** – the shelter or resting place of a wild animal.

**hydrophobic** – being resistant to water.

**mutant** – a person or thing that has changed in genetic form and qualities.

**prey** – animals hunted or killed by other animals for food.

**regenerating** – regrowing and replacing lost or destroyed tissue.

**reinforced** – strengthened by adding materials or support.

**solitary** – living alone.

Wolverine also has three retractable blade-claws in each hand. He releases them when threatened or in a fight. Wolverine, like his animal counterpart, is a lone fighter who you would want on your side!